Joseph R. Drake

The Culprit Fay

Joseph R. Drake

The Culprit Fay

ISBN/EAN: 9783743335080

Manufactured in Europe, USA, Canada, Australia, Japa

Cover: Foto ©berggeist007 / pixelio.de

Manufactured and distributed by brebook publishing software
(www.brebook.com)

Joseph R. Drake

The Culprit Fay

THE CULPRIT FAY.

· THE

CULPRIT FAY.

BY

JOSEPH RODMAN DRAKE.

NEW YORK:

RUDD & CARLETON, 130 GRAND STREET,

(BROOKS BUILDING, COR. OF BROADWAY.)

MDCCCLIX.

ADVERTISEMENT.

——o——

"The exquisite poem of 'THE CULPRIT FAY,' was composed hastily among the Highlands of the Hudson, in the summer of 1819. The author—says his biography—was walking with some friends on a warm moonlight evening, when one of the party remarked that it would be difficult to write a faery poem, purely imaginative, without the aid of human characters. When the party was reassembled, two or three days afterward, 'THE CULPRIT FAY' was read to them, nearly as it is now printed."

" My visual orbs are purged from film, and, lo!
        Instead of Anster's turnip-bearing vales
    I see old fairy land's miraculous show !
        Her trees of tinsel kissed by freakish gales,
    Her Ouphs that, cloaked in leaf-gold, skim the breeze,
        And fairies, swarming————————"

                                        TENNANT'S ANSTER FAIR.

POEM.

L

'Tıs the middle watch of a summer's
    night—
The earth is dark, but the heavens are
    bright;
Naught is seen in the vault on high
But the moon, and the stars, and the
    cloudless sky,
And the flood which rolls its milky hue,
A river of light on the welkin blue.
The moon looks down on old Cronest,
She mellows the shades, on his shaggy
    breast,
And seems his huge grey form to throw
In a silver cone on the wave below;

His sides are broken by spots of shade,
By the walnut bough and the cedar
    made,
And through their clustering branches
    dark
Glimmers and dies the fire-fly's spark—
Like starry twinkles that momently break,
Through the rifts of the gathering tem-
    pest's rack.

II.

The stars are on the moving stream,
    And fling, as its ripples gently flow,
A burnished length of wavy beam
    In an eel-like, spiral line below;
The winds are whist, and the owl is still,

The bat in the shelvy rock is hid.
And naught is heard on the lonely hill
But the cricket's chirp, and the answer
    shrill
    Of the gauze-winged katy-did;
And the plaint of the wailing whip-poor-
    will,
    Who moans unseen, and ceaseless sings,
Ever a note of wail and woe,
    Till morning spreads her rosy wings,
And earth and sky in her glances glow.

III.

'Tis the hour of fairy ban and spell;
The wood-tick has kept the minutes
    well;

He has counted them all with click and
　　stroke
Deep in the heart of the mountain oak,
And he has awakened the sentry elve
　Who sleeps with him in the haunted tree,
To bid him ring the hour of twelve,
　And call the fays to their revelry;
Twelve small strokes on his tinkling bell—
('Twas made of the white snail's pearly
　　shell;)
" Midnight comes, and all is well!
Hither, hither, wing your way!
'Tis the dawn of the fairy-day."

IV.

They come from beds of lichen green,
They creep from the mullen's velvet screen;

Some on the backs of beetles fly
From the silver tops of moon-touched trees,
 Where they swung in their cobweb ham-
  mocks high,
And rocked about in the evening breeze;
 Some from the hum-bird's downy nest—
They had driven him out by elfin power,
 And, pillowed on plumes of his rain-
  bow breast,
Had slumbered there till the charmed hour;
 Some had lain in the scoop of the rock,
With glittering ising-stars inlaid;
 And some had opened the four-o'clock,
And stole within its purple shade.
 And now they throng the moonlight glade,
Above—below—on every side,
 Their little minim forms arrayed
In the tricksy pomp of fairy pride!

V.

They come not now to print the lea,
In freak and dance around the tree,
Or at the mushroom board to sup,
And drink the dew from the butter-
    cup;
A scene of sorrow waits them now,
For an Ouphe has broken his vestal vow;
He has loved an earthly maid,
And left for her his woodland shade;
He has lain upon her lip of dew,
And sunned him in her eye of blue,
Fanned her cheek with his wing of air,
Played in the ringlets of her hair,
And, nestling on her snowy breast,
Forgot the lily-king's behest.
For this the shadowy tribes of air

To the elfin court must haste away :—
And now they stand expectant there,
  To hear the doom of the culprit Fay.

VI.

The throne was reared upon the grass,
Of spice-wood and of sassafras ;
On pillars of mottled tortoise-shell
  Hung the burnished canopy—
And over it gorgeous curtains fell
  Of the tulip's crimson drapery.
The monarch sat on his judgment-seat,
  On his brow the crown imperial shone,
The prisoner Fay was at his feet,
  And his peers were ranged around the
    throne,

He waved his sceptre in the air,
  He looked around and calmly spoke;
His brow was grave and his eye severe,
  But his voice in a softened accent
    proke :

VII.

" Fairy! Fairy! list and mark:
  Thou hast broke thine elfin chain;
Thy flame-wood lamp is quenched and
    dark,
  And thy wings are dyed with a deadly
    stain—
Thou hast sullied thine elfin purity
  In the glance of a mortal maiden's
    eye,

Thou hast scorned our dread decree,
　And thou shouldst pay the forfeit high,
But well I know her sinless mind
　Is pure as the angel forms above,
Gentle and meek, and chaste and kind,
　Such as a spirit well might love;
Fairy! had she spot or taint,
Bitter had been thy punishment.

Tied to the hornet's shardy wings;
Tossed on the pricks of nettles' stings;
Or seven long ages doomed to dwell
With the lazy worm in the walnut-
　　shell;
Or every night to writhe and bleed
Beneath the tread of the centipede;
Or bound in a cobweb dungeon dim,
Your jailer a spider huge and grim,

Amid the carrion bodies to lie,

Of the worm, and the bug, and the mur-
    dered fly:

These it had been your lot to bear,

Had a stain been found on the earthly
    fair.

Now list and mark our mild decree—

Fairy, this your doom must be:

VIII.

" Thou shalt seek the beach of sand

Where the water bounds the elfin
    land;

Thou shalt watch the oozy brine

Till the sturgeon leaps in the bright
    moonshine,

Then dart the glistening arch below,
And catch a drop from his silver bow.
The water-sprites will wield their arms
    And dash around, with roar and rave.
And vain are the woodland spirits' charms,
    They are the imps that rule the wave.
Yet trust thee in thy single might:
If thy heart be pure and thy spirit right,
Thou shalt win the warlock fight.

IX.

"If the spray-bead gem be won,
    The stain of thy wing is washed
        away:
But another errand must be done
    Ere thy crime be lost for aye;

Thy flame-wood lamp is quenched and
    dark,
Thou must reillume its spark.
Mount thy steed and spur him high
To the heaven's blue canopy;
And when thou seest a shooting star,
Follow it fast, and follow it far—
The last faint spark of its burning train
Shall light the elfin lamp again.
Thou hast heard our sentence, Fay;
Hence! to the water-side, away!"

x.

The goblin marked his monarch well;
    He spake not, but he bowed him low,
Then plucked a crimson colen-bell,
    And turned him round in act to go.

The way is long, he cannot fly,
  His soiled wing has lost its power,
And he winds adown the mountain high,
  For many a sore and weary hour.
Through dreary beds of tangled fern,
Through groves of nightshade dark and
    dern,
Over the grass and through the brake,
Where toils the ant and sleeps the
    snake;
  Now over the violet's azure flush
He skips along in lightsome mood;
  And now he thrids the bramble-bush,
Till its points are dyed in fairy blood.
He has leaped the bog, he has pierced
    the brier,
He has swum the brook, and waded the
    mire,

Till his spirits sank, and his limbs grew
   weak,
And the red waxed fainter in his cheek.
He had fallen to the ground outright,
   For rugged and dim was his onward
   track,
But there came a spotted toad in sight,
   And he laughed as he jumped upon
   her back:
He bridled her mouth with a silkweed
   twist,
He lashed her sides with an osier thong;
And now, through evening's dewy mist,
   With leap and spring they bound
   along,
Till the mountain's magic verge is past,
And the beach of sand is reached at
   last.

XI.

Soft and pale is the moony beam,
Moveless still the glassy stream;
The wave is clear, the beach is bright
 With snowy shells and sparkling stones;
The shore-surge comes in ripples light,
 In murmurings faint and distant moans;
And ever afar in the silence deep
Is heard the splash of the sturgeon's
  leap,
And the bend of his graceful bow is
  seen—
A glittering arch of silver sheen,
Spanning the wave of burnished blue,
And dripping with gems of the river-
  dew.

XII.

The elfin cast a glance around,
   As he lighted down from his courser
   toad,
Then round his breast his wings he
   wound,
   And close to the river's brink he strode;
He sprang on a rock, he breathed a
   prayer,
   Above his head his arms he threw,
Then tossed a tiny curve in air,
   And headlong plunged in the waters
   blue.

XIII.

Up sprung the spirits of the waves,
From the sea-silk beds in their coral caves,

With snail-plate armor snatched in haste,
They speed their way through the liquid
    waste;
Some are rapidly borne along
On the mailed shrimp or the prickly
    prong,
Some on the blood-red leeches glide,
Some on the stony star-fish ride,
Some on the back of the lancing squab,
Some on the sideling soldier-crab;
And some on the jellied quarl, that
    flings
At once a thousand streamy stings;
They cut the wave with the living
    oar,
And hurry on to the moonlight shore,
To guard their realms and chase away
The footsteps of the invading Fay.

XIV.

Fearlessly he skims along,
His hope is high, and his limbs are
    strong,
He spreads his arms like the swallow's
    wing,
And throws his feet with a frog-like
    fling;
His locks of gold on the waters shine,
    At his breast the tiny foam-bees rise,
His back gleams bright above the brine,
    And the wake-line foam behind him
    lies.
But the water-sprites are gathering near
    To check his course along the tide;
Their warriors come in swift career
    And hem him round on every side;

On his thigh the leech has fixed his
hold,

The quarl's long arms are round him
rolled,

The prickly prong has pierced his skin,

And the squab has thrown his javelin,

The gritty star has rubbed him raw,

And the crab has struck with his giant
claw;

He howls with rage, and he shrieks with
pain,

He strikes around, but his blows are
vain;

Hopeless is the unequal fight,

Fairy! naught is left but flight,

xv.

He turned him round, and fled amain
With hurry and dash to the beach again,
He twisted over from side to side,
And laid his cheek to the cleaving tide;
The strokes of his plunging arms are
    fleet,
And with all his might he flings his feet,
But the water-sprites are round him still,
To cross his path and work him ill.
They bade the wave before him rise;
They flung the sea-fire in his eyes,
And they stunned his ears with the scallop-
    stroke,
With the porpoise heave and the drum-
    fish croak.
Oh! but a weary wight was he

When he reached the foot of the dog-
    wood tree.
—Gashed and wounded, and stiff and sore,
He laid him down on the sandy shore;
He blessed the force of the charmed line,
    And he banned the water-goblin's spite,
For he saw around in the sweet moon-
    shine
Their little wee faces above the brine,
    Giggling and laughing with all their
    might
At the piteous hap of the Fairy wight.

XVI.

Soon he gathered the balsam dew
    From the sorrel-leaf and the henbane
    bud;

Over each wound the balm he drew,
  And with cobweb lint he stanched the
    blood.
The mild west wind was soft and low,
It cooled the heat of his burning brow,
And he felt new life in his sinews shoot,
As he drank the juice of the calamus
    root   .
And now he treads the fatal shore,
As fresh and vigorous as before.

XVII.

Wrapped in musing stands the sprite:
'Tis the middle wane of night;
  His task is hard, his way is far,
But he must do his errand right

Ere dawning mounts her beamy car,
And rolls her chariot wheels of light;
And vain are the spells of fairy-land,
He must work with a human hand.

XVIII.

He cast a saddened look around,
  But he felt new joy his bosom swell,
When, glittering on the shadowed ground,
  He saw a purple mussel-shell;
Thither he ran, and he bent him low,
He heaved at the stern and he heaved at
    the bow,
And he pushed her over the yielding sand,
Till he came to the verge of the haunted
    land.

She was as lovely a pleasure-boat
  As ever fairy had paddled in,
For she glowed with purple paint with-
    out,
  And shone with silvery pearl within;
A sculler's notch in the stern he made,
An oar he shaped of the bootle blade;
Then sprung to his seat with a lightsome
    leap,
And launched afar, on the calm, blue
    deep.

XIX.

The imps of the river yell and rave;
They had no power above the wave,
But they heaved the billow before the
    prow,

And they dashed the surge against her
side,

And they struck her keel with jerk and
blow,

Till the gunwale bent to the rocking
tide.

She wimpled about to the pale moon-
beam,

Like a feather that floats on a wind-
tossed stream ;

And momently athwart her track

The quarl upreared his island back,

And the fluttering scallop behind would
float,

And patter the water about the boat ;

But he bailed her out with his colen-bell,

And he kept her trimmed with a wary
tread,

While on every side like lightning fell
  The heavy strokes of his bootle-blade.

XX.

Onward still he held his way,
Till he came where the column of moon-
    shine lay,
And saw beneath the surface dim.
The brown-backed sturgeon slowly swim;
Around him were the goblin train—
But he sculled with all his might and
    main.
And followed wherever the sturgeon led,
Till he saw him upward point his head;
Then he dropped his paddle-blade,
And held his' colen-goblet up
To catch the drop in its crimson cup.

XXI.

With sweeping tail and quivering fin,
   Through the wave the sturgeon flew,
And, like the heaven-shot javelin,
   He sprung above the waters blue.
Instant as the star-fall light
   He plunged him in the deep again,
But left an arch of silver bright,
   The rainbow of the moony main.
It was a strange and lovely sight
   To see the puny goblin there;
He seemed an angel form of light,
   With azure wing and sunny hair,
   Throned on a cloud of purple fair,
Circled with blue and edged with white,
And sitting at the fall of even
Beneath the bow of summer heaven.

XXII.

A moment, and its lustre fell ;
 But ere it met the billow blue,
He caught within his crimson bell
 A droplet of its sparkling dew—
Joy to thee, Fay ! thy task is done,
Thy wings are pure, for the gem is won—
Cheerly ply thy dripping oar,
And haste away to the elfin shore.

XXIII.

He turns, and, lo ! on either side
The ripples on his path divide ;·
And the track o'er which his boat must
 pass
Is smooth as a sheet of polished glass.

Around, their limbs the sea-nymphs lave,
  With snowy arms half swelling out,
While on the glossed and gleamy wave
  Their sea-green ringlets loosely float;
They swim around with smile and song;
  They press the bark with pearly hand,
And gently urge her course along,
  Toward the beach of speckled sand;
  And, as he lightly leaped to land,
They bade adieu with nod and bow,
  Then gaily kissed each little hand,
And dropped in the crystal deep below.

XXIV.

A moment stayed the fairy there;
He kissed the beach and breathed a prayer;

Then spread his wings of gilded blue,
And on to the elfin court he flew;
As ever ye saw a bubble rise,
And shine with a thousand changing dyes,
Till, lessening far, through ether driven,
It mingles with the hues of heaven;
As, at the glimpse of morning pale,
The lance-fly spreads his silken sail,
And gleams with blendings soft and
    bright,
Till lost in the shades of fading night;
So rose from earth the lovely Fay--
So vanished, far in heaven away!

*   *   *   *   *   *   *

Up, Fairy! quit thy chick-weed bower,
The cricket has called the second hour,
Twice again, and the lark will rise
To kiss the streaking of the skies—

Up! thy charmed armor don,
Thou'lt need it ere the night be gone.

XXV.

He put his acorn helmet on;
It was plumed of the silk of the thistle-
    down;
The corslet plate that guarded his breast
Was once the wild bee's golden vest;
His cloak, of a thousand mingled dyes,
Was formed of the wings of butterflies;
His shield was the shell of a lady-bug
    queen,
Studs of gold on a ground of green;
And the quivering lance which he bran-
    dished bright,

Was the sting of a wasp he had slain in
   fight.
Swift he bestrode his fire-fly steed;
   He bared his blade of the bent grass
      blue;
He drove his spurs of the cockle-seed,
   And away like a glance of thought he
      flew,
To skim the heavens, and follow far
The fiery trail of the rocket-star.

XXVI.

The moth-fly, as he shot in air,
Crept under the leaf, and hid her there;
The katy-did forgot its lay,
The prowling gnat fled fast away,

The fell mosquito checked his drone,
And folded his wings till the Fay was
gone,
And the wily beetle dropped his head,
And fell on the ground as if he were
dead;
They crouched them close in the dark-
some shade,
They quaked all o'er with awe and
fear,
For they had felt the blue-bent blade,
And writhed at the prick of the elfin
spear;
Many a time, on a summer's night,
When the sky was clear and the moon
was bright,
They had been roused from the haunted
ground

By the yelp and bay of the fairy hound;
  They had heard the tiny bugle-horn,
They had heard the twang of the maize-
    silk string,
    When the vine-twig bows were tightly
      drawn,
    And the needle-shaft through air was
      borne,
Feathered with down of the hum-bird's
    wing.
And now they deemed the courier ouphe,
  Some hunter-sprite of the elfin ground;
And they watched till they saw him mount
    the roof
    That canopies the world around;
Then glad they left their covert lair,
And freaked about in the midnight air.

XXVII.

Up to the vaulted firmament
His path the fire-fly courser bent,
And at every gallop on the wind,
He flung a glittering spark behind;
He flies like a feather in the blast
Till the first light cloud in heaven is
    past.
  But the shapes of air have begun their
  work,
And a drizzly mist is round him cast,
  He can not see through the mantle
  murk,
He shivers with cold, but he urges fast;
  Through storm and darkness, sleet and
  shade,
He lashes his steed and spurs amain,

For shadowy hands have twitched the
   rein,
  And flame-shot tongues around him
    played,
And near him many a fiendish eye
Glared with a fell malignity,
And yells of rage, and shrieks of fear,
Came screaming on his startled ear.

XXVIII.

His wings are wet around his breast,
The plume hangs dripping from his crest,
His eyes are blurred by the lightning's
   glare,
And his ears are stunned with the thun-
   der's blare,

But he gave a shout, and his blade he
   drew,
  He thrust before and he struck behind,
Till he pierced their cloudy bodies through,
  And gashed their shadowy limbs of
   wind;
Howling the misty spectres flew,
  They rend the air with frightful cries,
For he has gained the welkin blue,
  And the land of clouds beneath him
   lies.

XXIX.

Up to the cope careering swift,
  In breathless motion fast,
Fleet as the swallow cuts the drift,

3

Or the sea-roc rides the blast,
The sapphire sheet of eve is shot,
   The sphered moon is past,
The earth but seems a tiny blot
   On a sheet of azure cast.
O! it was sweet, in the clear moonlight,
   To tread the starry plain of even,
To meet the thousand eyes of night,
   And feel the cooling breath of heaven!
But the elfin made no stop or stay
Till he came to the bank of the milky-
   way,
Then he checked his courser's foot,
And watched for the glimpse of the planet-
   shoot.

XXX.

Sudden along the snowy tide
    That swelled to meet their footsteps' fall,
The sylphs of heaven were seen to glide,
    Attired in sunset's crimson pall;
Around the Fay they weave the dance,
    They skip before him on the plain,
And one has taken his wasp-sting lance,
    And one upholds his bridle-rein;
With warblings wild they lead him on
    To where, through clouds of amber
    seen,
Studded with stars, resplendent shone
    The palace of the sylphid queen.
Its spiral columns, gleaming bright,
Were streamers of the northern light;
Its curtain's light and lovely flush

Was of the morning's rosy blush,
And the ceiling fair that rose aboon
The white and feathery fleece of noon.

XXXI.

But, O! how fair the shape that lay
    Beneath a rainbow bending bright;
She seemed to the entranced Fay
    The loveliest of the forms of light;
Her mantle was the purple rolled
    At twilight in the west afar;
'Twas tied with threads of dawning gold,
    And buttoned with a sparkling star.
Her face was like the lily roon
    That veils the vestal planet's hue;
Her eyes, two beamlets from the moon,

Set floating in the welkin blue.
Her hair is like the sunny beam,
And the diamond gems which round it
    gleam
Are the pure drops of dewy even
That ne'er have left their native heaven.

XXXII.

She raised her eyes to the wondering
    sprite,
    And they leaped with smiles, for well
    I ween
Never before in the bowers of light
    Had the form of an earthly Fay been
    seen.
Long she looked in his tiny face;

Long with his butterfly cloak she played;
She smoothed his wings of azure lace,
  And handled the tassel of his blade;
And as he told in accents low
The story of his love and woe,
She felt new pains in her bosom rise,
And the tear-drop started in her eyes.
And " O, sweet spirit of earth," she cried,
  "Return no more to your woodland
    height,
But ever here with me abide
  In the land of everlasting light!
Within the fleecy drift we'll lie,
  We'll hang upon the rainbow's rim;
And all the jewels of the sky
  Around thy brow shall brightly beam!
And thou shalt bathe thee in the stream
  That rolls its whitening foam aboon,

And ride upon the lightning's gleam,
 And dance upon the orbed moon!
We'll sit within the Pleiad ring,
 We'll rest on Orion's starry belt,
And I will bid my sylphs to sing
 The song that makes the dew-mist melt;
Their harps are of the umber shade,
 That hides the blush of waking day,
And every gleamy string is made
 Of silvery moonshine's lengthened ray;
And thou shalt pillow on my breast,
 While heavenly breathings float around,
And, with the sylphs of ether blest,
 Forget the joys of fairy ground."

XXXIII.

She was lovely and fair to see,
And the elfin's heart beat fitfully;
But lovelier far, and still more fair,
The earthly form imprinted there;
Naught he saw in the heavens above
Was half so dear as his mortal love,
For he thought upon her looks so meek,
And he thought of the light flush on her
        cheek ;
Never again might he bask and lie
On that sweet cheek and moonlight eye,
But in his dreams her form to see,
To clasp her in his revery,
To think upon his virgin bride,
Was worth all heaven, and earth beside.

XXXIV.

"Lady," he cried, " I have sworn to-night,
On the word of a fairy-knight,
To do my sentence-task aright;
My honor scarce is free from stain,
I may not soil its snows again;
Betide me weal, betide me wo,
Its mandate must be answered now."
Her bosom heaved with many a sigh,
The tear was in her drooping eye;
But she led him to the palace-gate,
     And called the sylphs who hovered
          there,
And bade them fly and bring him straight
     Of clouds condensed a sable car.
With charm and spell she blessed it there,
From all the fiends of upper air;

Then round him cast the shadowy shroud,
And tied his steed behind the cloud;
And pressed his hand as she bade him
    fly
Far to the verge of the northern sky,
For by its wane and wavering light
There was a star would fall to-night.

XXXV.

Borne afar on the wings of the blast,
Northward away, he speeds him fast,
And his courser follows the cloudy wain
Till the hoof-strokes fall like · pattering
    rain.
The clouds roll backward as he flies,
Each flickering star behind him lies,

And he has reached the northern plain,
And backed his fire-fly steed again,
Ready to follow in its flight
The streaming of the rocket-light.

XXXVI.

The star is yet in the vault of heaven,
   But it rocks in the summer gale;
And now 'tis fitful and uneven,
   And now 'tis deadly pale;
And now 'tis wrapped in sulphur-smoke,
   And quenched is its rayless beam,
And now with a rattling thunder-stroke
   It bursts in flash and flame.
As swift as the glance of the arrowy lance
   That the storm-spirit flings from high,

The star-shot flew o'er the welkin blue,
   As it fell from the sheeted sky.
As swift as the wind in its trail behind
   The elfin gallops along,
The fiends of the clouds are bellowing
   loud,
But the sylphid charm is strong;
He gallops unhurt in the shower of fire,
   While the cloud-fiends fly from the
   blaze;
He watches each flake till its sparks
   expire,
   And rides in the light of its rays.
But he drove his steed to the lightning's
   speed,
   And caught a glimmering spark;
Then wheeled around to the fairy ground,
   And sped through the midnight dark.

*   *   *   *   *   *   *

Ouphe and Goblin! Imp and Sprite!
    Elf of eve! and starry Fay!
Ye that love the moon's soft light,
    Hither, hither wend your way;
Twine ye in a jocund ring,
    Sing and trip it merrily,
Hand to hand, and wing to wing,
    Round the wild witch-hazel tree.

Hail the wanderer again
    With dance and song, and lute and lyre,
Pure his wing and strong his chain,
    And doubly bright his fairy fire.
Twine ye in an airy round,
    Brush the dew and print the lea;
Skip and gambol, hop and bound,
    Round the wild witch-hazel tree.

The beetle guards our holy ground,
  He flies about the haunted place,
And if mortal there be found,
  He hums in his ears and flaps his face;
The leaf-harp sounds our roundelay,
  The owlet's eyes our lanterns be;
Thus we sing, and dance, and play,
  Round the wild witch-hazel tree.

But, hark! from tower on tree-top high,
  The sentry-elf his call has made:
A streak is in the eastern sky,
  Shapes of moonlight! flit and fade!
The hill-tops gleam in morning's spring,
The sky-lark shakes his dappled wing,
The day-glimpse glimmers on the lawn,
The cock has crowed, and the Fays are
  gone.

www.ingramcontent.com/pod-product-compliance
Lightning Source LLC
Chambersburg PA
CBHW022010190326
41519CB00010B/1472